CRYSTALLINE SILICA

Published by the

Society for Mining, Metallurgy, and Exploration, Inc.

On the Cover

Top photograph of silica crystal courtesy Pennsylvania Glass Sand Corporation, Berkeley Springs, West Virginia.

Bottom photograph of a sand and gravel operation courtesy National Aggregates Association, Silver Springs, Maryland.

Society for Mining, Metallurgy, and Exploration, Inc.
P.O. Box 6252002
Littleton, CO, U.S.A. 80162-5002
800-763-3132

ISBN 0-87335-166-5

Contents

PREFACE vii

INTRODUCTION ix

CHAPTER 1 **UNDERSTANDING CRYSTALLINE SILICA** 1

What Is Silica? 1

    Silicon Is an Element 1

    Silica Is a Chemical Compound 2

    Silicates Are Compounds of Silicon and Oxygen Plus
        Other Elements 5

    Silicones Are Synthetic Compounds 5

What Is Meant by *Crystalline?* 6

    The Crystalline State 6

    The Noncrystalline State 6

    Focusing on Crystalline Silica 8

Crystalline Silica's Forms 8

The Natural Occurrences of Crystalline Silica 10

    Igneous Rocks 11

    Sedimentary Rocks 12

    Metamorphic Rocks 12

Crystalline Silica in the Mineral Industry Setting 12

The Many Uses of Crystalline Silica 12

    Manufacturing of Glass, Ceramics, and Fine China 15

    Construction 15

    Heavy Industry 18

High-Tech Applications   **19**
Synthetic Crystalline Silica   **19**
Conclusions   **20**

CHAPTER 2   **THE REGULATION OF CRYSTALLINE SILICA   21**
OSHA's Hazard Communication Standard   **22**
The IARC Evaluation Process   **23**
IARC Classification of Silica   **25**
Amended IARC Evaluation of Crystalline Silica
    as a Carcinogen (1996)   **26**
Regulatory Activities of Other Agencies   **26**
The Complexities of Measurement   **27**
Conclusions   **29**

CHAPTER 3   **SELECTED READINGS AND OTHER RESOURCES   31**
Recommended Readings   **31**
    Basic Information   **31**
    Technical Data   **31**
Other Resources   **32**

GLOSSARY   33

INDEX   43

. . . .

Figures and Tables

Figures

1.1   Relationships among chemical terms   **2**

1.2   Crystal structure of quartz   **3**

1.3   Photomicrographs of diatoms   **4**

1.4   Tiles in a repeating pattern   **7**

1.5   Tiles in a random pattern   **7**

1.6   Tiles in a nonrepeating pattern   **7**

1.7   Relationships among forms of silica as well as silicon, silicone, and silicate   **8**

1.8   Stability fields of the different forms of silica   **9**

1.9   The rock cycle   **10**

1.10   The glass-making process   **17**

1.11   The Smithsonian Castle   **17**

1.12   Cross section of an oil well bore   **18**

Tables

1.1   Silica in commodities and end-product applications   **13**

1.2   Common products containing 0.1% or more crystalline silica   **16**

2.1   Methods used to detect quartz in a sample   **28**

. . . .

Preface

This publication, *Crystalline Silica,* consists of the material covered in the general information booklet *Crystalline Silica Primer* (now out of print; originally published in 1992 by the U.S. Bureau of Mines, U.S. Department of the Interior) and a section that updates the discussion of the carcinogenicity of crystalline silica. The International Agency for Research on Cancer (IARC) in 1996 reevaluated the classification of crystalline silica to group 1—carcinogenic to humans. This decision, coupled with the demise of the U.S. Bureau of Mines, was the catalyst that led SME to publish this introductory book on crystalline silica, a substance so common in the mining field as well as everyday life that its dangers are often unrecognized.

The SME Foundation is proud to be a partner with SME in underwriting this publishing endeavor. It is important that SME members, the mining industry, and the general public have good information available to them for meeting the challenges of addressing the crystalline silica issue and its impact on the mining industry. The many donors to the SME Foundation have made this publication possible.

SME acknowledges the cooperation of the U.S. Geological Survey, particularly Aldo Barsotti and Robert L. Virta. Virta, with coauthor Sarkis G. Ampian, laid the foundation for *Crystalline Silica* in a technical paper, which is listed in the selected readings section.

. . . .

Introduction

Crystalline silica is the scientific name for a group of minerals composed of silicon and oxygen. The term *crystalline* refers to the fact that the silicon and oxygen atoms are arranged in a three-dimensional repeating pattern. This group of minerals has shaped human history since the beginning of civilization. From the sand used for making glass to the piezoelectric quartz crystals used in advanced communication systems, crystalline silica has been a part of our technological development. Crystalline silica's pervasiveness in our technology is matched only by its abundance in nature. It is found in rocks and sediments from every geologic era and from every continent around the globe.

Scientists have known for decades that prolonged and excessive exposure to crystalline silica dust in mining environments can cause silicosis, a noncancerous lung disease. During the 1980s, studies were conducted that suggested that crystalline silica also was a carcinogen. As a result of these findings, crystalline silica has been regulated under the Hazard Communication Standard (HCS) of the U.S. Occupational Safety and Health Administration (OSHA). Under the HCS, OSHA-regulated businesses that use materials containing 0.1% or more crystalline silica must follow federal guidelines concerning hazard communication and worker training. Although the HCS does not require that samples be analyzed for crystalline silica, mineral suppliers or OSHA-regulated businesses may choose to do so if they wish to show that they are exempt from the requirements of the HCS.

Because crystalline silica forms the extremely common mineral quartz as well as less common minerals and because the HCS affects production and use of many mineral commodities, it is important that there be as clear an understanding as possible of what is and what is not crystalline silica, where it is found and used, and how it is qualitatively and quantitatively identified. This publication addresses these issues in as nontechnical a manner as possible. Chapter 1 describes, in largely nonscientific terms, what crystalline silica is and how we come in contact with it. Chapter 2 discusses the regulatory decisions that have created new interest in this widespread substance and presents a nontechnical overview of the techniques used to determine the presence and abundance of crystalline silica. Because this publication is meant to be a starting point for anyone interested in learning more about crystalline silica, a list of selected readings and other resources is included. The detailed glossary, which defines many terms that are beyond the scope of this publication, is designed to help the reader move from this presentation to a more technical one, the inevitable next step.

CHAPTER 1

Understanding Crystalline Silica

Crystalline silica is an indispensable part of both the natural and the technological worlds. We all come in contact with it daily and have all our lives. It has been called one of the building blocks of our planet. Although it is a mainstay of modern technology, it is neither modern nor manufactured. It was known to the ancients, and its uses are still being expanded today.

WHAT IS SILICA?

Other substances whose names sound similar—silicon, silicone, and silicates—are sometimes confused with silica. The terms may sound alike, but each means something quite distinct. Knowing the differences among these four substances is crucial to understanding what crystalline silica is and, perhaps just as important, what it is not (Figure 1.1).

Silicon Is an Element

All matter in the universe is formed from the 107 or so chemical elements known to exist. A chemical element is a fundamental constituent of matter that consists of only one kind of atom. *Silicon* is the second most common element in the Earth's crust, second only to oxygen, and together silicon and oxygen make up approximately 75% of the Earth on which we live and from which we get all that we use in our daily lives.

Strictly speaking, silicon (whose chemical symbol is Si) is classified as a nonmetal, but it possesses some of the properties associated with metals. There are eight elements, in fact, that fall on the borderline between metals and

1

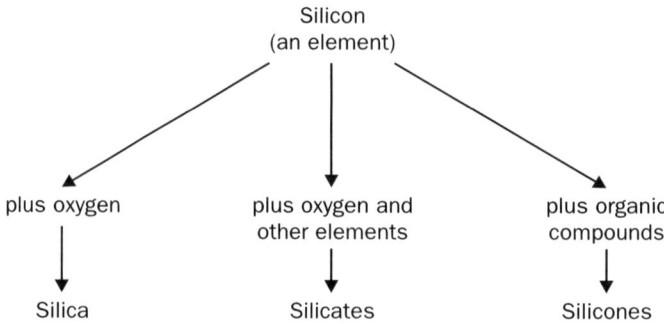

FIGURE 1.1   Relationships among chemical terms

nonmetals. Some scientists refer to these as *metalloids*. One property associated with metals, for example, is their ability to conduct electricity. Silicon's electronic capabilities are unusual: At high temperatures, it acts like a metal and conducts electricity, but at low temperatures, it acts like an insulator and does not. It is said to be a *semiconductor*. This unusual property made silicon the perfect element to move technology first into the world of transistors, then into the world of integrated circuits, and finally into the world of today's computer chips.

Silicon is the backbone of a computer chip. The pure silicon needed for this use, however, does not exist in nature; it is formed from silica sand (described next). Thin slices of pure silicon are then etched with the intricate electronic circuits needed to run a computer.

Silica Is a Chemical Compound

The compound *silica* (SiO_2) is formed from silicon and oxygen atoms (chemical symbol of oxygen is O). A chemical compound is defined as a distinct and pure substance formed by the union of two or more elements. Because oxygen is the most abundant element in the Earth's crust and silicon is the second most abundant, the formation of silica is quite common in nature. The silica sand, just mentioned as the substance used to derive pure silicon, is made of *quartz*, which is the most common form of silica found in nature. When quartz is studied on an atomic scale by using X-ray analysis, its *crystalline* structure can be deduced. Four oxygen atoms link together to form a three-dimensional shape called a tetrahedron. In the center of this structure is one silicon atom.

Therefore, the structure is called a *silicon-oxygen* (SiO$_4$) *tetrahedron. Tetrahedron* means "four surfaces" and refers to the shape of the SiO$_4$ structure. To form a quartz crystal, myriads of the three-dimensional tetrahedra are joined together by sharing one another's corner oxygens (Figure 1.2).

Silica can also be biological in origin, produced by tiny organisms. The most significant of these are *diatoms* (plants; Figure 1.3) and *radiolarians* (animals), both of which extract dissolved silica from the water around them and deposit it in an *amorphous* (i.e., *noncrystalline* or *cryptocrystalline*) state to form their cell walls. For both organisms, silica is a nutrient they must have to survive. In nature, they use the silica that dissolves out of rocks. When diatoms grown in the laboratory extract all the available silica from the aquarium

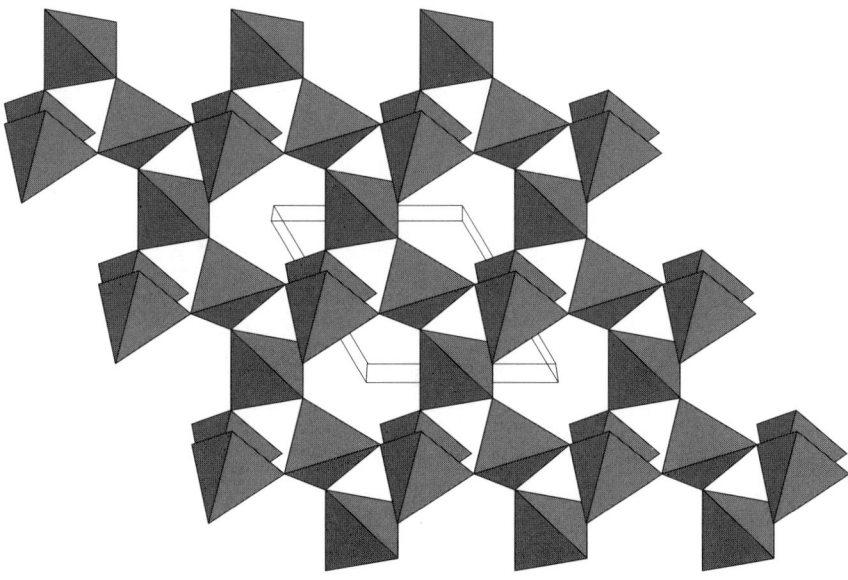

FIGURE 1.2 **Crystal structure of quartz. The triangular shapes are the sides of tetrahedra. Four triangles form one tetrahedron, which is a three-dimensional shape like a pyramid but with a triangular base instead of a square one. At each point of each tetrahedron is an oxygen atom (i.e., four oxygen atoms per tetrahedron). Inside each tetrahedron is a silicon atom. The tetrahedra link together by sharing the oxygen atoms and thus form a very strong structure that gives quartz its physical characteristics. [Courtesy Joseph R. Smyth, Department of Geological Sciences, University of Colorado, Boulder.]**

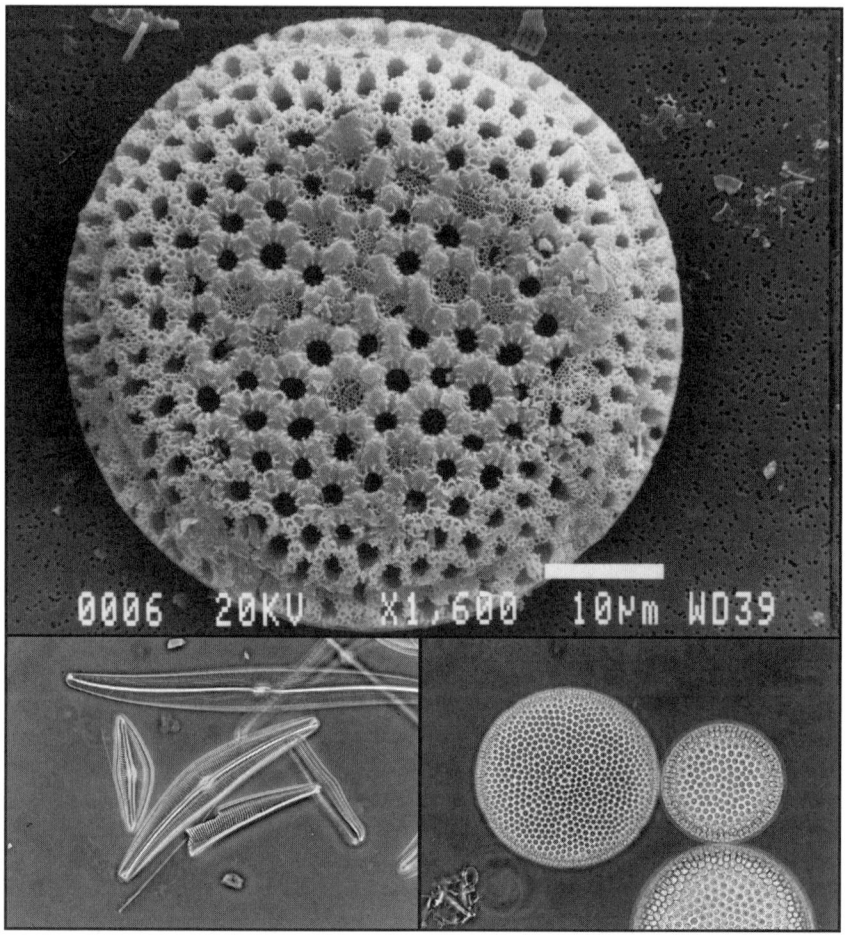

FIGURE 1.3 Photomicrographs of diatoms. Diatoms are microscopic single-celled organisms in the plant kingdom. They secrete silica to form cell walls. The enlargement on the top shows the delicate structure of the silica wall. The two photomicrographs on the bottom show three species. [Top: *Coscinodiscus* sp. diatom (diameter is approximately 53 µm or 0.002 inch; specimen is from Celite Corporation's diatomite mine, Lompoc, California); courtesy World Minerals, Santa Barbara, California. Bottom left: *Gyrosigma* sp. (upper) and *Cymbella* sp. (lower) diatoms; courtesy Rick Ellis. Bottom right: *Coscinodiscus* sp. diatom; courtesy Rick Ellis.]

water, they attach themselves to the walls of the aquarium and use small amounts of dissolved silica etched from the glass itself. In nature, when diatoms and radiolarians die, they sink to the bottom of the water and accumulate as a sediment, which can become hardened into diatomite and radiolarite. Diatomite is a commercially useful rock. It is highly porous and, thus, is effective for filtering as well as for use as a filler and as a mild abrasive.

Thus, silica can be found in more than one state—*amorphous* as in the remains from a diatom and *crystalline* as in a quartz crystal. This difference is further explained below. Both states of silica are composed of SiO_2, but they are quite different physically. What's more, silica in its crystalline state is found in more than one form. This phenomenon is called *polymorphism* (literally "many forms").

Silicates Are Compounds of Silicon and Oxygen Plus Other Elements

Even when bonding with other elements, silicon and oxygen remain together in *silicon-oxygen* (SiO_4) *tetrahedra.* The silicon-oxygen tetrahedra bond most commonly with sodium, potassium, calcium, magnesium, iron, and aluminum to form *silicates.*

Silicates constitute the most abundant class of minerals. Geologists regard silicate minerals as the basic material out of which most rocks are created.

Silicones Are Synthetic Compounds

Silicones are *polymers,* a type of synthetic compound. Developed commercially during World War II, silicones are formed from two or more silicon atoms linked with carbon compounds (referred to as *organic compounds)*. Most silicones contain oxygen as well. Unlike what happens when silica and silicates form, in silicone, the silicon and oxygen do not take the tetrahedral shape but instead form chainlike structures called *silicon polymers.* Polymerization is a chemical reaction in which small organic molecules combine to form larger molecules that contain repeating structural units of the original molecules.

Silicone materials range from liquids (used as water repellents and defoamers) to greases and waxes (used as water- and heat-resistant lubricants) to resins and solids (used to make special heat- and chemical-resistant products including paints, rubbers, and plastic parts). Probably silicone's most highly publicized use is in the manufacture of breast implants.

WHAT IS MEANT BY *CRYSTALLINE?*

We mentioned that the compound silica, which is formed by the chemical reaction of silicon and oxygen, can be either crystalline or noncrystalline. Depending upon the extremes of temperature and pressure that a solid compound has been subjected to or, in some cases, the speed at which it cooled, the solid can take on different forms. Diatomite, described earlier, and quartz are identical chemically (both are composed of SiO_2) and both are solids at room temperature, but their physical forms—and their internal structures—are very different.

The Crystalline State

In a crystalline substance (such as quartz), the atoms and molecules make up a three-dimensional repeating pattern. The pattern unit is repeated indefinitely in three directions, forming the crystalline structure. This process is similar to arranging floor tiles, in which a two-dimensional pattern unit—say, one made of two black tiles and two white tiles—is repeated indefinitely in two directions (Figure 1.4). This repeating pattern can be altered. It would be possible to change the positions of the two black tiles and two white tiles in relationship to one another and still have a pattern that could be repeated indefinitely in two directions, but the resulting design would be different. Likewise, the internal structure of the crystal can be changed, and the resulting crystalline substance would be changed.

The Noncrystalline State

Now, picture the black tiles and white tiles placed randomly on the floor, forming no pattern whatsoever (Figure 1.5). Such is the structure of a noncrystalline, or amorphous, substance. A diatom's cell wall is an example of silica in a noncrystalline state.

Some amorphous materials exhibit short-range ordering of their atoms. Using the analogy of the floor tiles one last time, suppose that the black and white tiles formed a pattern in some small regions, but it was a *nonrepeating* pattern (Figure 1.6). The distinguishing feature of a crystalline substance is that you can take any portion of it and see the whole. With a nonrepeating pattern, you can't do that. Some short-range orderliness may exist, but no predictable order extends over a long distance.

Scientists call this state *glassy.* Not surprisingly, window glass, which forms when molten glass is cooled very quickly, is an example of silica in a glassy

FIGURE 1.4　Tiles in a repeating pattern. In concept, this repeating pattern is similar to the repeating pattern of silica in crystalline state, as in quartz.

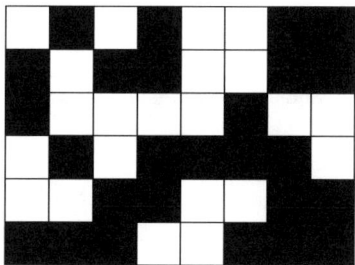

FIGURE 1.5　Tiles in a random pattern. In concept, this random pattern is similar to the random pattern of silica in an amorphous state, as in a diatom.

FIGURE 1.6　Tiles in a nonrepeating pattern. This pattern displays short-range order of the tiles, but long-range disorder. In concept, this nonrepeating pattern is similar to the nonrepeating pattern of silica in a glassy state, as in window glass.

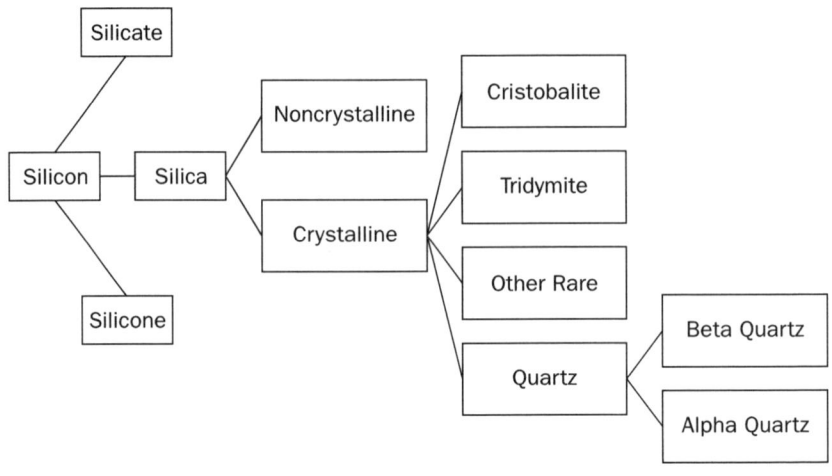

FIGURE 1.7   **Relationships among forms of silica as well as silicon, silicone, and silicate**

state. It is not crystalline because it cooled too rapidly—a process known as *quenching*—for the atoms to arrange themselves into a long-range periodically repeating structure, but it contains short-range ordering that many amorphous materials do not possess. Glassy and amorphous materials are considered to be synonymous by many scientists because both are noncrystalline.

Focusing on Crystalline Silica

We have discussed four terms that are often confused—silicon, silica, silicates, and silicones, and we have narrowed our discussion to silica, the compound formed from the elements silicon and oxygen.

We have seen how silica can be crystalline or noncrystalline. This publication's focus is on silica in its crystalline state only (Figure 1.7).

CRYSTALLINE SILICA'S FORMS

Crystalline silica exists in seven different forms or polymorphs, four of which are extremely rare. The three major forms—quartz, cristobalite, and tridymite—are stable at different temperatures (Figure 1.8). Within the three major

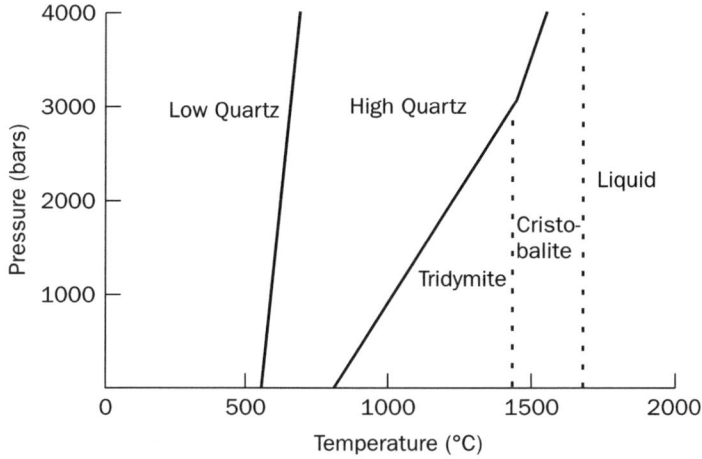

FIGURE 1.8 Stability fields of the different forms of silica. For reference, normal atmospheric pressure is about 1 bar, and 21 ℃ (70 ℉) is a comfortable temperature; the plot shows that all silica forms except low (or alpha) quartz require much-higher-than-normal temperatures for their formation. All these forms are encountered in nature. The higher-temperature forms are common in glass, ceramic, and fine china manufacturing.

forms, there are subdivisions. Geologists distinguish, for example, between alpha (low) and beta (high) quartz, noting that at 573°C (1063°F) and at normal room pressure (1 bar = 29.5 inches of mercury = normal atmospheric pressure), quartz changes from one form to the other (alpha to beta). Each of these subdivisions is stable under different thermal conditions. Foundry processes, the burning of waste materials, and other manufacturing procedures can create the kinds of conditions necessary for quartz to change form. In nature, quartz in its alpha, or low, form is most common, although both lightning strikes and meteorite impacts can change alpha quartz into *keatite* or *coesite*. Alpha quartz is abundant, found on every continent in large quantities. In fact, alpha quartz is so abundant and the other polymorphs of crystalline silica are so rare that some writers use the specific term *quartz* in place of the more general term *crystalline silica*.

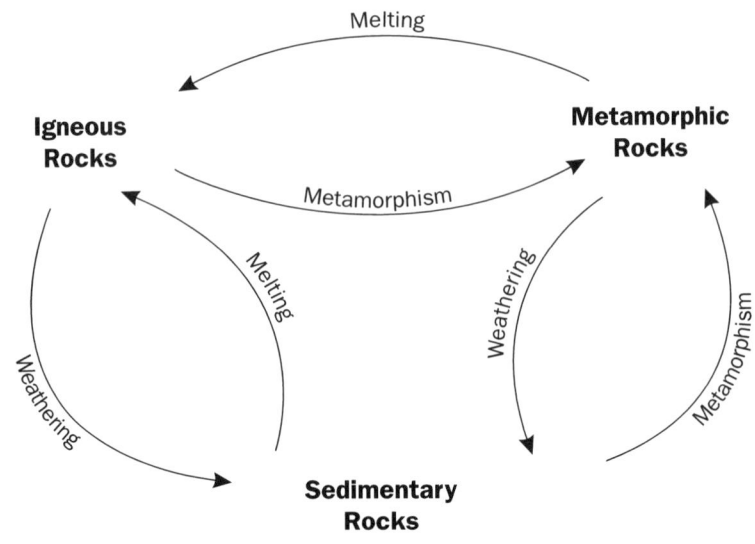

FIGURE 1.9   The rock cycle

THE NATURAL OCCURRENCES OF CRYSTALLINE SILICA

Quartz is one of the most common minerals. Besides occurring in rocks, crystalline silica in the form of quartz is a component of soil. It may have been part of the rock that weathered to form the soil, it may have been transported, or it may have crystallized from initially amorphous (that is, noncrystalline) silica that formed during the weathering process. Quartz is also the major component of most sand and of dust in the air.

Quartz is found in all three of the types of rocks that occur on Earth. *Igneous* rocks originate from magma, the material carried to the surface from the Earth's partially molten mantle (the region between the Earth's crust and its core). The other two types of rocks are *sedimentary* and *metamorphic*. Quartz is abundant in most examples of all three types of rock. It is one of the Earth's primary building blocks.

The *rock cycle* describes the relationships among the three types of rock (Figure 1.9). Igneous rocks reflect the activity of heat and pressure within the Earth that leads to melting; metamorphic rocks reflect the activity of heat and pressure that changes rocks without melting them; and sedimentary rocks reflect conditions of erosion and deposition (by wind, water, and ice) at the Earth's surface. Over geologic time, sedimentary rocks may be altered by heat and/or pressure to create metamorphic or igneous rocks. All rocks may be eroded to make sediments that, in turn, harden (*lithify*) into sedimentary rocks. Thus, the history of rocks that form the Earth's crust is one of continuous change. During these changes, quartz often endures. It is one of Earth's harder minerals, so it resists erosion, and it is soluble in very few chemicals, so it is seldom dissolved.

Igneous Rocks

Quartz is present in igneous rocks—but only those that contain excess silica. As magma cools, other minerals including olivine, pyroxenes, amphiboles, feldspars, and micas crystallize first. All of these minerals are silicates and need silicon and oxygen to form, because silicates are made from silicon, oxygen, and a metallic element, usually one of the six most common metallic elements. Quartz forms only if sufficient silicon and oxygen are left over after these silicates have formed. Nature's odds are stacked in quartz's favor, however. The fact that quartz is the second most common mineral in the Earth's crust (feldspar is most common) indicates that plenty of silicon and oxygen is usually left over after the crystallization of the other minerals to allow ample quantities of quartz to form. In fact, the average quartz content of igneous rocks is 12%.

Crystalline silica is present in those types of igneous rocks that contain excess silica. It is a common component of granite, rhyolite, quartz diorite, quartz monzonite, and andesite, to name a few. Crystalline silica as quartz also may be present in deposits of hardened, or *consolidated*, volcanic ash, known as *volcanic tuff*. When magma spews from a volcano, it drops in temperature so rapidly that the rock formed is usually glassy, that is, in a noncrystalline state. The 1980 eruption of Mount St. Helens is a perfect example of this process. If, however, the silica crystallizes before the magma leaves the volcano, then the resulting quartz appears as crystals surrounded by a glassy matrix. Volcanic glasses do crystallize over time, so a complex mixture of finely crystalline quartz and silicates eventually replaces the volcanic glass. Cristobalite and tridymite, the rarer, high-temperature-crystallizing forms of crystalline silica, may also be present in volcanic tuffs.

Sedimentary Rocks

Crystalline silica in the form of quartz is an extremely common component of sedimentary rocks. Sedimentary rocks form when minerals released during weathering or by chemical precipitation accumulate in a basin and are consolidated. Quartz, which is extremely resistant to physical and chemical breakdown by the weathering process, stays intact chemically even when fragmented and dispersed by erosion, wind, or other weathering processes. Quartz is present in a variety of sedimentary rock types, ranging from sandstones to conglomerates, in trace to major amounts.

Metamorphic Rocks

Metamorphic rocks, which form through heat or pressure intense enough to change the original rock without melting it, also contain crystalline silica as quartz. New textures may be created in the rock (for example, lineations may form or crystal sizes may increase), and new minerals may be formed during metamorphism. Quartz may be present in the original rock, it may crystallize from silica-bearing fluids that entered the rock during metamorphism, or it may form from a breakdown of other preexisting minerals as part of the metamorphic transformation.

CRYSTALLINE SILICA IN THE MINERAL INDUSTRY SETTING

Because of its abundance in the Earth, silica, in both its crystalline and noncrystalline states, is present in nearly all mining operations. It is in the host rock, in the ore being mined, as well as in what geologists call the overburden—the soil and surface material above the bedrock. Most ores are mined from deposits containing crystalline silica. The minerals forming the deposit and, to some extent, the processing of the ore determine the quartz content of the final product. Sand and gravel often consist mostly of quartz, whereas the quartz content of crushed stone varies from region to region. Table 1.1 lists some common commodities, the form of silica they contain, and their commercial uses.

THE MANY USES OF CRYSTALLINE SILICA

Crystalline silica, again primarily in the form of quartz, has been mined for thousands of years. In the first century A.D., the Roman scholar Pliny described the formation of quartz in some detail, although his understanding

TABLE 1.1   Silica in commodities and end-product applications

Commodity	Form of Silica in the Deposit	Major Commercial Applications
Antimony	Quartz	Flame retardants, batteries, ceramics, glass, alloys
Bauxite	Quartz	Aluminum production, refractories, abrasives
Beryllium	Quartz	Electronic applications
Cadmium	Quartz, jasper, opal, agate, chalcedony	Batteries, coatings and platings, pigments, plastics, alloys
Cement	None	Concrete (quartz in concrete mix)
Clay	Quartz, cristobalite	Paper, ceramics, paint, refractories
Copper	Quartz	Electrical conduction, plumbing, machinery
Crushed stone	Quartz	Construction
Diatomite	Quartz, amorphous silica	Filtration aids
Dimension stone	Quartz	Building facings
Feldspar	Quartz	Glass, ceramics, filler material
Fluorspar	Quartz	Acids, steel-making flux, glass, enamel, weld rod coatings
Garnet	Quartz	Abrasives, filtration, gemstone
Germanium	Quartz, jasper, opal, agate, chalcedony	Infrared optics, fiber optics, semiconductors
Gold	Quartz, chert	Jewelry, dental, industrial, monetary
Gypsum	Quartz	Gypsum board (prefabricated building product), industrial and building plaster
Industrial sand	Quartz	Glass, foundry sand
Iron ore	Chert, quartz	Iron and steel industry
Iron oxide pigment (natural)	Chert, quartz, amorphous silica	Construction materials, paint, coatings
Lithium	Quartz	Ceramics, glass, aluminum production
Magnesite	Quartz	Refractories

(Continues)

(Continued) TABLE 1.1   Silica in commodities and end-product applications

Commodity	Form of Silica in the Deposit	Major Commercial Applications
Mercury	Quartz	Chlorine and caustic soda manufacture, batteries
Mica	Quartz	Joint cement, paint, roofing
Perlite	Amorphous silica, quartz	Building construction products
Phosphate rock	Quartz, chert	Fertilizers
Pumice	Volcanic glass, quartz	Concrete aggregate, building block
Pyrophyllite	Quartz	Ceramics, refractories
Sand and gravel	Quartz	Construction
Selenium	Quartz	Photocopiers, glass manufacturing, pigments
Silicon	Quartz	Silicon and ferrosilicon for ferrous foundry and steel industry; computers; photoelectric cells
Silver	Quartz, chert	Photographic material, electrical and electronic products
Talc	Quartz	Ceramics, paint, plastics, paper
Tellurium	Quartz	Steel and copper alloys, rubber compounding, electronics
Thallium	Quartz, jasper, opal, agate, chalcedony	Electronics, superconductors, glass alloys
Titanium	Quartz	Pigments for paint, paper, plastics, metal for aircraft, chemical processing equipment
Tungsten	Quartz	Cemented carbides for metal machining and wear-resistant components
Vanadium	Quartz, amorphous silica	Alloying element in iron, steel, and titanium
Zinc	Quartz, jasper, opal, agate, chalcedony	Galvanizing, zinc-based alloys, chemicals, agriculture
Zircon	Quartz	Ceramics, refractories, zirconia production

was limited by the technology of his time. The ancients believed quartz to be very deep frozen ice, which could no longer be remelted. In fact, to prove this hypothesis, Pliny pointed out that quartz seemed to be found most frequently in the vicinity of glaciers. Although they may not have understood its true nature, early civilizations did understand its value as a gemstone, especially its purple form called amethyst. Today, quartz is used for a whole spectrum of products (Table 1.2) from high-technology applications in the electronics and optical fields to everyday uses in building and construction.

Manufacturing of Glass, Ceramics, and Fine China

One of the major uses of crystalline silica is as a raw material for glass manufacture. The first glass was probably made in Egypt more than 5,000 years ago. Today, the process has become highly refined (Figure 1.10). To ensure a very pure product, the specifications for glass are exceptionally stringent. A pure crystalline silica is used; the iron content must be less than 0.03%, and there are strict limits on the amounts of other impurities. Even the grain size of the crystals is specified. In the finished glass, the silica content must be at least 98.5%. Ceramics, porcelain, and fine china are made from finely ground crystalline silica, called *silica flour.*

Construction

Building materials, such as dimension stone (slabs and blocks of sandstone, granite, and limestone are examples) and concrete, contain crystalline silica in the form of quartz. Dimension stone is commonly used to build churches, government buildings, and monuments. In Washington, D.C., for example, the White House is built of sandstone, the Smithsonian Institution's original building—the Smithsonian Castle (Figure 1.11)—is sandstone, the exterior of the Museum of Natural History is granite, and the Treasury building is granite and sandstone. Quartz is a component of cement, another technological development dating from ancient times. In the past, sandpaper and grinding wheels were made from quartz, and it was the primary abrasive used in sandblasting operations.

Quartz is also used as functional filler in plastics, rubber, and paint. In George Washington's time, it was the fashion to add sand to paint. Thus, the wooden exterior of Mount Vernon, Washington's home in Virginia, was painted with a sand-paint mixture to give it the look of stone.

TABLE 1.2   Common products containing 0.1% or more crystalline silica

At Work	At Home	Everywhere
(exposure can occur in the process of manufacturing or using the following items)	(exposure can occur to a consumer of the following items)	(exposure can occur on the job or at home)
Asphalt filler is usually composed of quartz and stone aggregate.	Art clays and glazes contain clay and, sometimes, crystalline silica.	Caulk and putty contain clay, which can have a low to moderate crystalline silica content, as a filler.
Bricks have a high concentration of sand; they contain quartz and possibly cristobalite.	Cleansers contain pumice and feldspar as abrasives.	Dust (whether household or industrial) contains crystalline silica.
Concrete, like asphalt filler, contains stone aggregate.	Cosmetics contain talc and clay (both are silicates).	Fill dirt and topsoil contain sand. Because the crystalline silica content of common soil is so high, agricultural workers represent the occupational group most at risk for exposure to respirable crystalline silica.
Jeweler's rouge contains cryptocrystalline silica.	Pet litter is composed primarily of clay.	
Amethyst and quartz are crystalline silica materials that are used as jewelry.	Talcum powder contains talc (a silicate).	
Mortar contains sand.	Unwashed root vegetables (such as potatoes and carrots) are coated with soil, which has a high crystalline silica content.	Foam in furniture and on rug backings contains talc (a silicate) and silica.
Municipal water filter beds are constructed from both sand (crystalline silica) and diatomite (amorphous silica).	Pharmaceuticals contain clays and talc as fillers (both are silicates). Often the dosage of active ingredient in a medication is so minute that filler (listed as an inert ingredient) must be added to make the substance manageable to take.	Paint contains clay, talc, sand, and diatomite.
Plaster is made from gypsum but sometimes contains silica.		Paper and paper dust contain clay.
Plastic in appliances can contain clay, talc, crushed limestone, and silica as fillers.		
Roofing granules are made from sand and aggregate.	Sand is crystalline silica. Beach sand, play sand for sandboxes, and the sand used on golf courses are no different than industrial sand used for construction, in sandblasting, or on icy roads. All are largely crystalline silica.	
Wallboard is made from gypsum.		

FIGURE 1.10   The glass-making process [Courtesy U.S. Silica Company, Berkeley Springs, West Virginia.]

FIGURE 1.11   The Smithsonian Castle. The exterior is built of sandstone. [Courtesy Smithsonian Institution.]

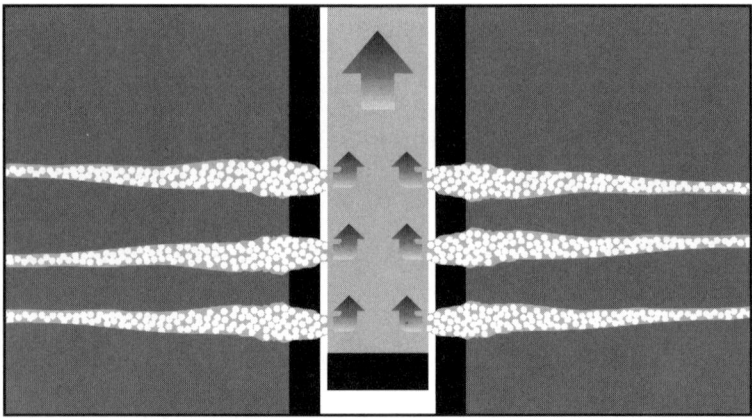

FIGURE 1.12   Cross section of an oil well bore. The initially solid rock is fractured by a pressure-driven water-sand mixture. The sand (shown by fine dots) remains in the fractures to hold them open. Oil (and gas, if present) flows through the spaces between the sand grains in the fractures and into and up the well bore to the surface. [Courtesy Unimin Corporation, New Canaan, Connecticut.]

Heavy Industry

Foundry molds and cores for the production of metal castings are made from quartz sand. The manufacture of silica brick that is used in the linings of glass- and steel-melting furnaces because it does not melt at their high temperatures of operation represents another common use of crystalline silica in industry.

The oil and gas industry uses crystalline silica to help break up rock in wells. The operator pumps a water-sand mixture, under pressure, into the rock formations to fracture them so that oil and gas may flow into the well bore and thence up to the surface (Figure 1.12). The sand remains in the fractures to hold them open; otherwise the weight of the overlying rock would quickly close the fractures. More than 1 million tons of quartz sand were used annually for this purpose during the 1970s and early 1980s when oil-well drilling was at its peak.

Quartz sand is also used for filtering sediment and bacteria from water supplies and in sewage treatment. Although this use of crystalline silica has increased in recent years, it still represents a small proportion of the total use.

High-Tech Applications

Historically, crystalline silica, as quartz, has been a material of strategic importance. During World War II, communications components in telephones and mobile military radios were made from quartz. With today's emphasis on military command, control, and communications surveillance and with modern advances in sophisticated electronic systems, quartz-crystal devices are in even greater demand.

In the field of optics, quartz meets many needs. It has certain optical properties that permit its use in polarized laser beams. The field of laser optics uses quartz as windows, prisms, optical filters, and timing devices. Smaller pieces of high-quality quartz crystals are used for prisms and lenses in optical instruments. Scientists are experimenting with quartz bars to focus sunlight in solar-power applications.

Quartz crystals possess a unique property called *piezoelectricity*. A piezoelectric crystal converts mechanical pressure into electricity and vice versa. When a quartz crystal is cut at an exact angle to its long axis, pressure on the crystal generates a minute electrical voltage, and likewise, an AC (alternating current) voltage applied to quartz causes it to vibrate more than 30,000 times per second in some applications.

Piezoelectric quartz crystals are used to make electronic oscillators, which provide accurate frequency control for radio transmitters and radio-frequency telephone circuits. Incoming signals of interfering frequencies can be filtered out by piezoelectric crystals. Piezoelectric crystals are also used for quartz watches and clocks and other time-keeping devices such as the crystal-controlled oscillators in computers.

Synthetic Crystalline Silica

Today, industry does not depend entirely on natural quartz for strategic applications. Since the 1940s, well-established techniques for synthetically growing quartz have been used and refined. Synthetic quartz crystals are grown in heavy-duty pressure cookers called *autoclaves* under pressures ranging from 10 to 138 MPa (1,500 to 20,000 pounds per square inch) and at temperatures of 250 to 450°C (480 to 840°F).

Natural crystalline silica may contain impurities or be flawed in some way, but synthetic crystals can be flawless. They can also be made to grow in a particular shape and size for specific needs.

CONCLUSIONS

Crystalline silica is the scientific name for a group of minerals composed of silicon and oxygen atoms that are arranged in a three-dimensional repeating pattern. The most common of the several forms of crystalline silica are quartz, cristobalite, and tridymite. Crystalline silica is ubiquitous, being in geologic materials from every geologic era and from every continent.

Crystalline silica's abundance in nature is matched by its pervasiveness in our technology. From the sand used for making glass to the piezoelectric quartz crystals used in advanced communication systems, crystalline silica has been a part of our technological development.

CHAPTER 2

The Regulation of Crystalline Silica

It has been estimated that over 3 million workers in the United States are exposed to silica dust in their occupations. Because of this contact there has been much interest in how exposure to crystalline silica affects one's health.

Scientists have known for decades that prolonged and excessive exposure to crystalline silica dust causes silicosis, a noncancerous lung disease. This association was most dramatically demonstrated by the significant increase in reported silicosis cases following the invention of the pneumatic hammer drill in 1897, the introduction of sandblasting in 1904, and the undertaking of a major tunneling project in the mid-1930s through a ridge of nearly pure quartz. These events helped spawn dust-control standards that have continued to evolve through the years.

During the 1980s, studies were conducted that suggested that crystalline silica also was a carcinogen. In 1987, the International Agency for Research on Cancer (IARC), an agency of the World Health Organization, evaluated the available medical literature on silica. On the basis of this evidence, IARC concluded that crystalline silica (but not noncrystalline, or amorphous, silica) was a group *2A substance,* a probable carcinogen for humans. Subsequent evaluation by IARC (discussed below in the section titled Amended IARC Evaluation of Crystalline Silica as a Carcinogen [1996]) led to the conclusion that crystalline silica if inhaled should be considered carcinogenic (i.e., a group 1 substance). This revised conclusion is discussed after the description of the method by which crystalline silica was assigned to group 2A.

The conclusions in the 1980s that crystalline silica was probably carcinogenic attracted attention for several reasons. With the publication of the first study suggesting that crystalline silica was a carcinogen, crystalline silica became regulated under the Hazard Communication Standard (HCS) of the U.S. Occupational Safety and Health Administration (OSHA) as a carcinogen. Under the HCS, OSHA-regulated businesses that use materials containing 0.1% or more crystalline silica must follow federal HCS guidelines concerning hazard communication and worker training. Although the IARC finding did not trigger any federal regulations, the work of that organization is important because it has captured the attention of the mining industry. The following sections discuss OSHA's HCS and the IARC evaluation process.

OSHA'S HAZARD COMMUNICATION STANDARD

In the development of the HCS, OSHA realized that the task before it was herculean: to evaluate all the substances to which workers are exposed, as many as 650,000 of which were potentially hazardous. In addition, once a substance had been determined to be hazardous, OSHA's rule-making process was time-consuming. OSHA decided that little information about hazards and protective measures would be made available to employees if the substance-by-substance approach to analysis and regulation was the only one pursued. The agency decided to adopt a generic approach and promulgated the HCS, which requires container labeling, material safety data sheets, and training. Specifically, HCS has the following requirements:

- *Chemical manufacturers and importers* must determine the hazards of the product. The regulation states that "if a mixture has been tested as a whole to determine its hazards, the results of such testing shall be used to determine whether the mixture is hazardous. If a mixture has not been tested as a whole to determine whether the mixture is a health hazard, the mixture shall be assumed to present the same health hazards as do the components that comprise 1% (by weight or volume) or greater of the mixture, except that the mixture shall be assumed to present a carcinogenic hazard if it contains a component in concentrations of 0.1% or greater which is considered to be a carcinogen under paragraph (d)(4) of this section."

- *Chemical manufacturers, importers, and distributors* must communicate the hazard information and associated protective measures to customers through the use of labels and material safety data sheets.

- *Employers* must (1) identify and list hazardous chemicals in their workplaces; (2) obtain material safety data sheets and labels for each hazardous chemical; (3) develop and implement a written hazard communication program, including labels, material safety data sheets, and employee training; and (4) communicate hazard information to their employees through labels, material safety data sheets, and formal training programs.

Application of the HCS to mixtures is based on the amount of the hazardous or carcinogenic material present. A mixture is considered to represent the same health hazard as any hazardous component present in concentrations of 1% or greater (by weight or volume) or any carcinogenic component present in concentrations of 0.1% or greater. Suppliers are held responsible for determining whether a substance is covered by OSHA's HCS, including whether the quantity of the hazardous chemical in a mixture exceeds these cutoffs. Testing is not required; the employer may assume that if the hazardous chemical is present, the mixture is covered by the HCS. The employer must label all hazardous chemicals in accordance with the HCS, provide material safety data sheets, and train exposed workers.

To help ensure that hazard evaluations by different suppliers would be consistent, OSHA refers to a number of existing documents as providing definitive findings of hazard. These include IARC and the National Toxicology Program (NTP) carcinogenicity reviews. In addition, hazardous and carcinogenic chemicals listed in the Code of Federal Regulations (29 CFR 1910, Subpart Z, *Toxic and Hazardous Substances,* OSHA) and in *Threshold Limit Values for Chemical Substances and Physical Agents in Work Environment* (American Conference of Governmental Industrial Hygienists) are covered under the HCS. However, one properly conducted scientific study in the literature establishes a hazard for purposes of the HCS.

THE IARC EVALUATION PROCESS

The International Agency for Research on Cancer, headquartered in Lyon, France, began its program to evaluate chemicals and cancer risks in 1967. It holds working group meetings in Lyon, usually one week long each, about three times a year. Each group focuses on a chemical or group of chemicals. Recently, other areas of study, such as radiation and viruses, have been included in addition to chemicals. Topics for study are selected on the basis of two criteria: evidence of human exposure *and* some evidence or suspicion of carcinogenicity.

IARC assigns an expert on the chemical to survey the scientific literature, review toxicology studies, and summarize the results. Assignments to subject-matter experts are usually made one year in advance of the working group meetings. During the week-long meeting, the working group breaks into two subgroups: chemistry and toxicology. These subgroups examine the validity of the materials supplied by the assigned experts. The two subgroups also agree on terminology and any special definitions required. At the end of the week, the two subgroups come together for a plenary session; the monographs are issued from the plenary session. If significant new data become available after a monograph has been published, the chemical will be reevaluated at a subsequent meeting, and a revised monograph may be published.

IARC does not commission any health studies or replicate any of the studies it uses, nor does it control the literature reviewed by the experts for its monograph assignments (although research may be sponsored by other IARC groups). The choice of studies to be included in the monograph is at the discretion of the expert panel.

IARC classifies the chemicals studied into four categories, as detailed next.

- *Group 1* The agent is carcinogenic to humans. This classification is reserved for those substances for which sufficient evidence of carcinogenicity in humans has been found. *Sufficient evidence in humans* is defined by IARC to mean a causal relationship between exposure to the agent and the development of human cancer.

- *Group 2* This group is divided into two subgroups: probably carcinogenic and possibly carcinogenic.

- *Group 2A* The agent is probably carcinogenic to humans. This classification is applied when studies demonstrate that there is limited evidence of carcinogenicity in humans and sufficient evidence of carcinogenicity in experimental animals. *Limited evidence in humans* is defined by IARC as evidence that shows some association between exposure to the agent and the development of cancer. The evidence is considered limited, however, because even though the two positives may be credible, chance, bias, or confounding factors cannot be ruled out. IARC defines *sufficient evidence in experimental animals* as documentation of a causal relationship based on the results from studies involving two or more species of animals or from two or more independent studies in one species carried out at different times or in different laboratories. In its Supplement 7 to *IARC Monographs on the Evaluation of Carcinogenic Risks to Humans,* IARC states, "in the absence of adequate data on humans, it is biologically plausible and prudent to regard agents for which there is *sufficient evidence of* carcinogenicity in experimental animals as if they presented a carcinogenic risk to humans" (p. 28).

- *Group 2B*   The agent is possibly carcinogenic to humans. When there is limited evidence in humans, but an absence of sufficient evidence in experimental animals, the chemical is classified as 2B.

- *Group 3*   The agent is not classifiable as to its carcinogenicity to humans. When studies do not provide sufficient data to classify a chemical into any of the other categories, IARC assigns it to group 3.

- *Group 4*   The agent is probably not carcinogenic to humans. IARC reserves this category for chemicals for which evidence from both human and experimental animal studies suggests a lack of carcinogenicity.

IARC CLASSIFICATION OF SILICA

IARC Monograph Volume 42 (1987) evaluates silica as follows:

- There is *sufficient evidence* for the carcinogenicity of crystalline silica to experimental animals.

- There is *inadequate evidence* for the carcinogenicity of amorphous silica to experimental animals.

- There is *limited evidence* for the carcinogenicity of crystalline silica to humans.

- There is *inadequate evidence* for the carcinogenicity of amorphous silica to humans.

Thus, when the working group reviewed the medical and scientific literature submitted to it for study, it found that there was evidence of an increased incidence of malignant tumors in animals exposed to crystalline silica. The working group also found a causal relationship in humans, although other confounding factors could not be excluded. These are the criteria required by IARC to classify something as a 2A substance, and crystalline silica was so classified. (See the next section for an update on this classification; crystalline silica when inhaled is now classified as a group 1 substance, i.e., it is thought to be carcinogenic to humans.)

It is important to note that not all the studies IARC examined show a link between exposure to silica and cancer. One important group of studies, cited in IARC's list of references, is compiled in *Silica, Silicosis, and Cancer: Controversy in Occupational Medicine* (1986), edited by David F. Goldsmith, Deborah M. Winn, and Carl M. Shy. Conclusions from several of the studies point to the controversy referred to in the volume's title:

- "Regulation of silica on the basis of potential carcinogenicity is premature" (p. 477, I.T.T. Higgins, "Is the Current Silica Standard Adequate?").

- "The present epidemiological and experimental data do not permit the conclusion that exposure to crystalline or amorphous SiO_2 is associated with increased risk of lung cancer" (p. 491, E. Mastromatteo, "Silica, Silicosis, and Cancer: A Viewpoint from a Physician Employed in Industry").

- "In reviewing the information we have on the health hazards of silica exposure, we find we have received positive and negative findings on the issue of silica and cancer, and our staff have told us a lot about interactions. They have told us that they are not certain it is silica alone that may be causing cancer" (p. 529, M. Schneiderman and D.M. Winn, "Where Are We with the SiO_2 and Cancer Issue?").

The following section provides an update on IARC activities concerning crystalline silica since the 1992 publication of the *Crystalline Silica Primer*.

AMENDED IARC EVALUATION OF CRYSTALLINE SILICA AS A CARCINOGEN (1996)

In October 1996, an IARC panel met to evaluate the carcinogenicity to humans from exposure to crystalline silica. The panel concluded that crystalline silica inhaled in the form of quartz or cristobalite from occupational sources should be classified as carcinogenic to humans (group 1). The change in classification was based on "a relatively large number of epidemiological studies that together provided sufficient evidence . . . for the carcinogenicity [in humans] of inhaled crystalline silica under the conditions specified." The panel found many cases of elevated lung cancer risk not explained by confounding factors. Rodent carcinogenicity studies supported the human evidence. The panel also found that "evidence that amorphous silica is a carcinogenic risk factor was considered to be inadequate upon grounds of both epidemiology and experimental studies. Amorphous silica was not classifiable as to its carcinogenicity to humans (Group 3)." The results of the IARC study are published in *Silica, Some Silicates, Coal Dust and para-Aramid Fibrils* in *IARC Monographs on the Evaluation of Carcinogenic Risks to Humans* (Volume 68).

REGULATORY ACTIVITIES OF OTHER AGENCIES

The Mine Safety and Health Administration (MSHA) and state legislators also are involved in the silica controversy. MSHA has proposed an HCS similar to OSHA's. MSHA's proposal also would use the results of IARC, NTP, and OSHA studies to determine which materials were carcinogenic. The enactment of such a regulation would affect nearly all mines because most ores are extracted from silica-bearing rock types. California passed the Safe Drinking Water and Toxic Enforcement Act of 1986, which includes crystalline silica of respirable size on its list of carcinogens. California's Air Toxic Hot Spots Act and Air Quality Act have the potential to restrict the emissions of crystalline silica. There is concern that burning of rice hulls and straw after harvesting generates an ash that usually contains residual biogenic silica. These regulations would have an adverse effect on the economics of these crops.

THE COMPLEXITIES OF MEASUREMENT

Crystalline silica is a very common material. It is present in most mineral operations, many of which sell crude or processed materials to OSHA-regulated sites. Sampling at factories, construction sites, or even the mine site will be greatly increased if a mineral supplier or OSHA-regulated employer wishes to prove that the concentration of crystalline silica is less than 0.1% for carcinogens, thereby exempting the company from the requirements of the HCS.

Because crystalline silica is a common mineral, it is assumed that it is easy to measure. Unfortunately that isn't always the case. Determining whether a particular *state* of silica is present and how much of that state a sample contains, however, can be a much more complicated problem. Under certain conditions, current techniques and equipment can't distinguish very well between silica's physical states at the low concentration specified by the HCS. Analysis can be difficult or even impossible for some samples. From an analyst's point of view, the problem is mainly one of equipment sensitivity.

In short, these are the problems analysts face when trying to so precisely pinpoint the crystalline silica content of a sample:

- *Variability*   The crystallinity of silica from different deposits, even from slightly different locations within the same deposits, is not necessarily the same. This variability raises two problems. First, a single standard (that is, the reference material to which the silica in the sample is compared) may not be appropriate. Using a standard that matches the particle size and crystallinity of the silica in the sample is essential for an accurate analysis. Second, obtaining a representative sample, when the sample size is so small and the deposit is so large, is nearly impossible.

TABLE 2.1   Methods used to detect quartz in a sample

Name	Description of Technique	Accuracy	Remarks
Optical microscopy	Samples are visually examined and the mineralogy is determined.	Accurate to within a few percent.	Requires considerable skill by analyst to identify the minerals present. Uses small samples.
Electron microscopy	Particle composition and morphology are determined. Crystal structure is determined with transmission electron microscopy. Resolves very small particles.	Accuracy limited by the nature of the analysis.	Cannot differentiate crystalline and amorphous silica except when transmission electron microscopy is used. Methods are slow and expensive; samples are very small.
Thermal analysis	Measures a mineral's response to temperature changes.	Accurate only for quantities over 1%.	Can be used only on very small samples.
Selective dissolution	Minerals are dissolved selectively by using acids. Quartz generally is less soluble than other minerals so it remains in the residue. Residue is analyzed to determine the content of crystalline silica.	Not very accurate.	Particle size and sample composition affect the accuracy of this method. Fine-grained quartz, cristobalite, and tridymite may dissolve; other minerals may not dissolve.
Separation based on density	A finely ground sample is suspended in a heavy liquid. The denser minerals settle faster than less dense minerals. By varying the density of the liquid, minerals with different densities can be separated from one another.	Not satisfactory for routine analysis.	Particle size, shape, and surface charge affect settling rates. The technique is slow and difficult to perform. Many of the heavy liquids used are highly toxic.
Infrared spectroscopy	Minerals absorb infrared light at specific wavelengths. By examining how the light is absorbed by the sample, the analyst can identify the minerals in the samples.	Accurate to about 1%.	Requires very small samples; the analyst must be sure that samples are representative of the deposit.
X-ray diffraction	X-rays are diffracted by the lattice planes of the minerals in the sample. By observing the intensity of the diffracted X-rays at different angles of incidence, the analyst can determine the identity and concentrations of minerals in sample.	Most accurate; typically, the limit is about 1%.	The degree of crystallinity (from amorphous to highly crystalline) and the presence of silicates can affect the accuracy of the quantitative analysis.

- *Presence of Other Minerals*   Nearly all samples contain more than just silica. Some of these minerals interfere with the data interpretation, making the accurate determination of silica content difficult or impossible.

- *Presence of Other States*   Some samples contain silica in more than one state. Because of the way it forms, silica may exhibit different degrees of crystallinity. Quartz can be intermingled with noncrystalline silica especially because natural quartz may have a noncrystalline coating. Silica that is cementing quartz grains together in a sandstone may have a different crystallinity than the quartz grains. Even X-ray diffraction, the most effective measurement technique, can distinguish between crystalline and noncrystalline silica only when rigid sampling protocol is followed and only when the sample has few other minerals.

- *Effect of Human Interference*   The very act of taking a sample can change the sample. The grinding and pulverizing required to get a small enough portion to analyze subjects the silica to heat and pressure, the very forces that, in nature, change crystalline surfaces to disordered amorphous surfaces. The analyst can't measure how much of it transforms from crystalline to noncrystalline during sample preparation.

The seven techniques that scientists use to measure how much crystalline silica a given substance contains are summarized in Table 2.1.

CONCLUSIONS

Scientists have known for decades that prolonged and excessive exposure to crystalline silica dust in mining environments can cause silicosis, a noncancerous lung disease. During the 1980s, studies suggested that crystalline silica was a carcinogen. In 1987, IARC labeled crystalline silica as a "2A substance," a probable human carcinogen, following a review of the available medical literature on silica. Subsequently, crystalline silica was upgraded to a group 1 substance (a human carcinogen) following a review of the health literature in 1996. With the publication of the first study, crystalline silica was regulated under OSHA's HCS as a carcinogen. To demonstrate that they are exempt from the requirements of HCS, suppliers must now analyze the crystalline silica content at the 0.1% level and must now more carefully consider whether the silica is crystalline or noncrystalline, whether it is a regulated form of crystalline silica, or whether it is a mixture of several silica types.

Because this action has implications that go beyond the mining and mineral-processing industries, it is important that there be as clear an understanding as possible about crystalline silica. This publication is meant to be a starting point from which to learn about crystalline silica's mineralogy, occurrences, and uses in society and about regulations affecting its use. A detailed glossary and a list of selected readings and other resources are included to help the reader move from this presentation to a more technical one, the inevitable next step.

CHAPTER 3

Selected Readings and Other Resources

RECOMMENDED READINGS

Basic Information

Boegel, H. 1968. *The Studio Handbook of Minerals.* Edited by J. Sinkankas. New York: Viking Press.

Miles, W.J. 1990. The Mining Industry Responds to Crystalline Silica Regulations. *Mining Engineering,* 19: 345–348.

Symes, R.F., and R.R. Harding. 1991. *Crystal and Gem.* New York: Alfred A. Knopf.

Zim, H.S., and P.R. Shaffer. 1957. *Rocks and Minerals.* New York: Western.

Technical Data

Ampian, S.G., and R.L. Virta. 1992. *Crystalline Silica Overview: Occurrence and Analysis.* U.S. Bureau of Mines Information Circular 9317.

Craighead, J.E., and the Silicosis and Silicate Disease Committee. 1988. Diseases Associated with Exposure to Silica and Nonfibrous Silicate Minerals. *Archives of Pathology and Laboratory Medicine,* 112: 673–720.

Hamilton, R.D., N.G. Peletis, and W.J. Miles. 1990. Detection and Measurement of Crystalline Silica in Minerals and Chemicals. In *Regulation of Crystalline Silica.* Littleton, Colorado: Manville Corporation.

Murray, H.H. 1990. *Occurrence and Uses of Silica and Siliceous Materials.* Littleton, Colorado: Society for Mining, Metallurgy, and Exploration, Inc. [Preprint]

World Health Organization, International Agency for Research on Cancer. 1987. *Silica and Some Silicates.* Lyon, France: *IARC Monographs on the Evaluation of the Carcinogenic Risk of Chemicals to Humans,* Volume 42.
World Health Organization, International Agency for Research on Cancer. 1996. *Silica, Some Silicates, Coal Dust and para-Aramid Fibrils.* Lyon, France: *IARC Monographs on the Evaluation of the Carcinogenic Risks to Humans,* Volume 68.

OTHER RESOURCES

Chemical Manufacturers' Association, Chemstar Crystalline Silica Panel, 2501 M Street, N.W., Washington, D.C. 20037
International Diatomite Producers Association, 26 Wind Jammer Court, Long Beach, California 90803
Refractories Institute, 500 Wood Street, Suite 326, Pittsburgh, Pennsylvania 15222
U.S. Geological Survey, Department of the Interior, 12201 Sunrise Valley Drive, Reston, Virginia 20192

Glossary

2a substance – A substance ranked by IARC as probably carcinogenic to humans, that is, there is limited evidence of carcinogenicity in humans and sufficient evidence of carcinogenicity in experiments with animals. (See *IARC Monographs on the Evaluation of Carcinogenic Risks to Humans.*) Crystalline silica was ranked as a 2A substance until 1996, when an IARC panel reevaluated its carcinogenicity and found sufficient evidence of human carcinogenicity to change the ranking to group 1.

agate – *Cryptocrystalline silica*. Composed of extremely fine (submicroscopic) crystals of silica.

aggregate – Either a group of materials or any of several hard, inert substances (such as sand, gravel, or crushed stone) used for mixing with cement.

amorphous – *See* **noncrystalline**.

atom – A minute particle of matter. The smallest particle of an element that can enter into chemical reactions.

bedrock – The rock underlying the soil or other surface material.

carcinogen – A substance that causes cancer. In scientific literature, the terms *tumorigen, oncogen*, and *blastomogenal* all have been used synonymously with carcinogen although occasionally, *tumorigen* has been used specifically to connote a substance that induces benign tumors. The *Federal Register* (Volume 52, Number 163, p. 31884) reports the following definition of *carcinogen* under federal regulation 29 CFR 1201:

A chemical is considered to be a carcinogen if:

(a) It has been evaluated by the International Agency for Research on Cancer (IARC), and found to be a carcinogen or potential carcinogen; or

(b) It is listed as a carcinogen or potential carcinogen in the Annual Report on Carcinogens *published by the National Toxicology Program (NTP) (latest edition); or*

(c) It is regulated by OSHA as a carcinogen.

chalcedony – *Cryptocrystalline silica.* Composed of extremely fine (submicroscopic) silica crystals.

chemical carcinogenesis – An IARC term, although one that is widely used elsewhere. It refers to the "induction by chemicals (or complex mixtures of chemicals) of *neoplasms* that are usually observed, the earlier induction of neoplasms that are commonly observed, and/or the induction of more neoplasms than are usually found" (IARC Monograph Volume 42).

chemical compound – A distinct and pure substance formed by the union of two or more elements in a definite proportion by weight.

chemical element – A fundamental substance that consists of only one kind of atom.

chert – *Cryptocrystalline silica.* Composed of extremely fine (submicroscopic) silica crystals.

coesite – A rare form of crystalline silica. Believed at first to exist only as a synthetic form of crystalline silica, it was formed in the laboratory by using two different methods. Subsequently, it has been found in nature at Meteor Crater in Arizona. It was also found to occur as a result of shock-wave experiments and nuclear explosions.

colloidal silica – Extremely fine amorphous silica particles dispersed in water. Colloids do not settle out of suspension over time. Colloidal silica is used commercially as a binder and stiffener and as a polishing agent.

compound – *See* **chemical compound**.

consolidation – In geological terms, any process by which loose, soft, or liquid Earth materials harden into rock.

cristobalite – The form of crystalline silica that is stable at the highest geologically attainable temperature. It occurs naturally in volcanic rock.

cryptocrystalline silica – Silica with submicrometer (i.e., less than 0.000001 m [= 1 μm = 0.00004 inch] across) crystals formed from amorphous, often biogenic, silica that undergoes compaction over geologic time. Examples are flint and chert. Also called *microcrystalline silica*.

crystalline – Having a highly structured molecular arrangement. The atoms and molecules form a three-dimensional, repeating pattern, or lattice.

derived – Refers to a substance formed from the products that result when a more complex substance is destroyed. Pure silicon is said to be derived from quartz sand.

devitrify – To change from glassy form to the crystalline state.

diatomaceous earth – *See* **diatomite**.

diatomite – A rock high in amorphous silica content, formed from the structures of tiny fresh- and salt-water organisms called diatoms. Diatomite has several commercial uses.

dimension stone – Building stone quarried and prepared in regularly shaped blocks to fit a particular design.

displacive transformation – A crystal transformation in which the atoms move slightly but no molecular bonds are broken. For example, the transformation of low quartz to high quartz involves only the rotation of the silicon-oxygen tetrahedra. Displacive transformations are usually rapid.

element – *See* **chemical element**.

epidemiological studies – Studies of illness involving human subjects over the long term. They generally involve analyses of real-world incidence of the illness with little or no attempt to control factors that potentially could contribute to the onset or severity of the illness. This approach contrasts with laboratory studies, which are generally performed on animals, are short term, and have variables that are controllable.

ethyl silicate, $Si(OC_2H_5)_4$ – A colorless, flammable liquid with a faint odor. It is an OSHA-regulated substance.

free silica – Informal name for pure crystalline silica that is chemically uncombined.

fused quartz – The material formed by the rapid melting of quartz crystals. A meteor strike or a lightning bolt striking sand can form fused quartz. The term *quartz glass* is often erroneously used to mean fused quartz, but quartz glass is a misnomer because quartz is crystalline and glass is noncrystalline.

fused silica – The material formed by heating cristobalite to the melting point (1710°C or 3110°F) and cooling it rapidly.

glassy – Having a semi-ordered molecular arrangement. Atoms and molecules may form a pattern, but it has only short-term or partial order and does not repeat predictably in three dimensions.

gravimetric sampling – Quantitative chemical sampling in which the substances in a compound are measured by weight.

host rock – A rock that contains ores or minerals of value.

IARC – The International Agency for Research on Cancer, an agency of the United Nations' World Health Organization.

igneous rock – A rock that has solidified from a molten state.

jasper – *Cryptocrystalline silica.* Composed of extremely fine (submicroscopic) silica crystals.

keatite – A synthetic and rare form of crystalline silica, formed by the crystallization of amorphous silica. It is transformed to cristobalite at a temperature of 1620°C (2948°F).

lithify – To harden into rock.

macrophage – A large phagocyte, a type of cell in the body that engulfs foreign materials and consumes debris and foreign bodies.

magma – A natural hot melt of rock-forming materials (primarily silicates) and steam. Magma is often in motion through the Earth's crust. Geologists speculate that certain magmas originate within the Earth's crust and others come from greater depths.

material impairment – Cited in the *Federal Register* (1/19/89), regulation 29 CFR 1910, as "life threatening effects; disabling effects; various diseases;

irritation to different organs or tissues; and changes in organ functions indicative of future health decrements" (p. 2361).

metal – A type of element. Metals are usually hard and lustrous, malleable (they can be pounded into sheets), and ductile (they can be drawn into wires) and can conduct electricity and heat.

metalloid – A group of elements, eight in all, that form the boundary (on the periodic table) between the metal elements (such as copper, iron, tin, gold) and the nonmetal elements (such as carbon, nitrogen, hydrogen). Metalloids possess some of the properties of the metals and some of the properties of the nonmetals.

metamorphic rock – Rocks that have undergone changes during exposure to the high pressures and temperatures in the Earth's interior.

metastable – Possessing an energy state that is not stable, yet will not change spontaneously. An outside force is required to change the energy state. *See also* **stable**.

methylsilicate – AN OSHA-regulated substance, $(CH_3O)_4Si$; it exists in the form of colorless needles.

microcrystalline silica – *See* **cryptocrystalline silica**.

mineral – Naturally occurring crystalline solids, most of which are made from oxygen, silicon, sulfur, and any of six common metals or metal compounds.

molecule – The smallest particle of a substance that retains the qualities of the substance and is composed of one or more atoms.

neoplasm – A tumor.

noncrystalline – Having an unstructured molecular arrangement. The atoms and molecules are randomly linked, forming no pattern.

nonmetal – Elements that do not exhibit the properties of metals. Usually poor conductors of electricity and heat.

opal – An amorphous form of silica.

organic compound – A chemical compound containing carbon.

OSHA – The Occupational Safety and Health Administration, an agency of the U.S. Department of Labor.

overburden – Material overlying the ore in a deposit.

permissible exposure limit (PEL) – An OSHA term. It refers to the concentration of a substance to which a worker is allowed to be exposed as a time-weighted average.

phagocytize – To remove from the body by the action of phagocytes, cells in the body that engulf foreign materials and consume debris and foreign bodies. It is believed that upon exposure to airborne crystalline silica particles, 80% of the particles are phagocytized and eliminated within a short time.

piezoelectricity – The ability of some crystals to convert mechanical pressure to electricity and to convert electricity to vibration. A quartz crystal in a watch is an example of applied piezoelectricity.

polymerization – A chemical reaction in which small organic molecules combine to form larger molecules that contain repeating structural units of the original molecules. The product of polymerization is called a *polymer*.

polymorph – Literally "many forms." To be polymorphic means to have or assume several forms. In reference to crystals, it is the characteristic of crystallizing in more than one form. For example, crystalline silica can be in the form of quartz, cristobalite, tridymite, or others.

precipitated silica – Amorphous silica that is precipitated from either a vapor or solution.

quartz – The most common type of crystalline silica. Some publications will use *quartz* and *crystalline silica* interchangeably, but the term *crystalline silica* actually encompasses several forms: quartz, **cristobalite, tridymite,** and several rarer forms.

radiolarian earth – Soil, high in amorphous silica content, composed predominantly of the remains of radiolaria. Radiolarian earth that has been consolidated (hardened) into rock is called **radiolarite**.

radiolarite – A rock, high in amorphous silica content, formed from the skeletons of tiny fresh- and salt-water organisms called radiolaria.

reconstructive transformation – A crystal transformation that involves the breaking of molecular bonds. For example, the transformation of quartz to tridymite involves the restructuring of the molecules. It is generally a slow transformation.

respirable crystalline silica (respirable dust) – May be defined as dust that contains particles small enough to enter the gas-exchange region of the human lung (about 3.5 μm [0.00014 inch]). One of the studies to which IARC refers in its monograph (Volume 42, 1987) found that the particle sizes of crystalline silica in a crushed alpha quartz sandstone product (with the trademark Min-U-Sil 15) were distributed as follows: particles larger than 5 μm (0.0002 inch) constituted about 0.1% of the sample, particles between 2 and 4.9 μm (between 0.00008 and 0.0002 inch), about 7%; and particles less than 1.9 μm (0.00007 inch), 92.8%. The Silicosis and Silicate Disease Committee (National Institute for Occupational Safety and Health) states that particles less than 1 μm (0.00004 inch) in size are the most troublesome and that particles in the range of 0.5 to 0.7 μm (0.00002 to 0.00003 inch) are retained in the lung (see *Archives of Pathology and Laboratory Medicine,* Volume 112, July 1988). As early as 1943, however, the Department of Labor established a limit of no more than 5 million particles of free silica under 10 μm (0.0004 inch) in size per 0.03 cubic meter (1 cubic foot) of air (see *Silicosis,* Industrial Health Series Number 9, U.S. Department of Labor, Division of Labor Standards, 1943).

rock cycle – A cycle taking place over geologic time in which the three types of rocks are transformed from one to another. *Sedimentary rocks* or *igneous rocks* are changed by heat and pressure into *metamorphic rocks* (or metamorphic rocks are changed again into different metamorphic rocks). Sedimentary, metamorphic, or igneous rocks are melted to create igneous rocks. Sedimentary, metamorphic, and igneous rocks are eroded to make sediments that then harden *(lithify)* into sedimentary rocks.

sedimentary rock – A rock formed by the accumulation and consolidation of minerals that have been either transported to a particular site by wind, water, or ice or precipitated by a chemical reaction at the site.

semiconductors – Materials that act as conductors within certain temperature ranges; at other temperatures they act as insulators. The elements silicon and germanium are examples of semiconductors of electricity. This unusual electrical capability has led to silicon's use in transistors, integrated circuits, and computer chips.

silica – A compound formed from silicon and oxygen. Silica is a polymorph, that is, it exists in more than one state. The states of silica are crystalline and noncrystalline (also called amorphous). Crystalline silica can take several forms: quartz (most common), cristobalite, tridymite, and four more rare forms.

silica brick – Brick composed of silica that is used as a lining in furnaces.

silica flour – Finely ground quartz. Typically 98% of the particles are smaller than 5 µm (0.0002 inch) in diameter.

silica gel – Amorphous silica, prepared in water. Removal of the liquid creates xerogels and further treatment with alcohol creates aerogels. Silica gels are used as drying agents and to alter viscosity of liquids.

silica sand – A common term in industry. It generally is used to mean a sand that has a very high percentage of silica, usually in the form of quartz. Silica sand is used as a source of pure silicon and as a raw material for glass and other products. Also called *quartz sand*.

silica W – A synthetic form of crystalline silica. It reacts rapidly with water, transforming into amorphous silica.

silicate minerals – Minerals containing silicon, oxygen, and a metal or metal compound. Silica tetrahedra form the framework of silicate minerals. Examples are olivine, pyroxene, amphibole, feldspar, and mica.

silicates – Compounds formed from silicon, oxygen, and other elements. *See also* **silicate minerals** *and* **silicon-oxygen tetrahedron**.

siliceous – A term used to describe a rock with a high silica content, especially one containing free silica rather than silicates.

silicic rock – An igneous rock containing more than two-thirds SiO_2, by weight, usually as quartz or feldspar. Granite is an example of silicic rock.

silicon – The second most common element in the Earth's crust. (Oxygen is the most common.) Silicon's chemical symbol is Si. Silicon is a metalloid, possessing some of the properties of a metal and some of the properties of a nonmetal. Pure silicon does not exist in nature. Silicon derived in the laboratory exists as black to gray, lustrous, needlelike crystals and is an OSHA-regulated substance.

silicon carbide (SiC) – A green to blue-black iridescent crystal. It is an OSHA-regulated substance.

silicon dioxide – Silica (SiO_2).

silicon-oxygen tetrahedron – Silicon and oxygen bonded in a set formation with four oxygen atoms and one silicon atom (see Figure 1.2). Its chemical symbol is SiO_4. *Tetrahedron* literally means "four surfaces" and refers to the way the molecule looks internally. Picture four spheres (the oxygen atoms) touching a smaller sphere (the silicon atom) held in the pocket in the middle of the oxygen spheres. Lines drawn between the centers of the oxygen spheres would form a regular, four-sided prism, a tetrahedron. Although many structures are possible in nature, geologists seldom encounter more than a relatively small number, primarily because most rocks are made up of silicate minerals, which combine silicon-oxygen tetrahedra with other elements. The silicon-oxygen tetrahedron bonds most frequently with sodium, potassium, calcium, magnesium, iron, and aluminum.

silicon tetrahydride (SiH_4) – A colorless gas used in the manufacture of semiconductors. Also called silane. It is an OSHA-regulated substance.

silicones – Synthetic compounds formed from two or more silicon atoms linked with carbon compounds. Most silicones contain oxygen as well. Silicones are formed by a process called polymerization; the molecular structure is a chain, not the tetrahedral shape of the molecules of silica or the silicates. *See also* **polymerization**.

silicosis – A lung disease characterized by scarring of lung tissue, which is contracted by prolonged exposure to high levels of respirable silica dust or acute levels of respirable silica dust.

stable – Possessing an energy state that is balanced and will not change spontaneously, resistant to energy change. *See also* **metastable**.

states of matter – A substance can be in a solid, liquid, or gas state. These three are called *states of matter*.

stishovite – The most dense form of crystalline silica. It is rare and at first was believed to exist only as a synthetic. It was initially found to occur as a result of shock-wave experiments and of nuclear explosions. Subsequently, it was found in nature at Meteor Crater in Arizona.

tetrahedron – A solid geometric shape with four surfaces (plural = tetrahedra). *See also* **silicon-oxygen tetrahedron**.

tridymite – A form of crystalline silica. It is found in nature in volcanic rocks and stony meteorites. It is also found in fired silica bricks.

vitreous silica – Glassy silica. The term is sometimes used to refer to any noncrystalline substance.

vitrify – To form as a glass.

volcanic tuff – Deposits of volcanic ash that have hardened into rock.

Index

Note: *f* indicates figure, *t* indicates table.

A
Agate 33
Aggregate 33
Air Quality Act (California) 27
Air Toxic Hot Spots Act (California) 27
Alpha quartz 8*f*, 9, 9*f*
Amethyst 15
Amorphous state 3, 5, 8. *See also*
   Glassy state
Atom 33
Autoclaves 19

B
Bedrock 33
Beta quartz 8*f*, 9, 9*f*

C
California
   legislation affecting silica 27
Carcinogens 33–34
Carcinogenic qualities vii, 21–23, 29
  IARC amended evaluation of
     silica 26
  IARC categories 24–25
  IARC classification of silica 25–26
Cement 15
Ceramics 15
Chalcedony 34
Chemical carginogenesis 34
Chemical compound 34
Chemical element 34
Chert 34

Coesite 9, 34
Colloidal silica 34
Consolidation 34
Construction materials 15
Cristobalite 8*f*, 11, 20, 34
   temperature stability 8–9, 9*f*
Cryptocrystalline silica 35
Crystalline silica 5, 8, 8*f*
   as carcinogen vii, 21–26, 28
   in ceramics 15
   in commodities 13*t*–14*t*
   in construction materials 15
   defined vii, 20, 40
   in dust 10
   in fine china 15
   forms 8–9
   in glass 15, 17*f*
   in igneous rocks 10–11
   in industry 18
   measurement variables and
     techniques 27, 29, 28*t*
   in metamorphic rocks 10
   in mining operations 12, 13*t*–14*t*
   natural occurrences 10–11
   polymorphism 5
   in porcelain 15
   products containing 15–19, 16*t*
   and quartz vii
   in rocks 10–11
   in sand 10
   in sedimentary rocks 10
   and silicosis vii, 21, 28, 41

Crystalline silica, *continued*
  in soil 10
  synthetic 19
  and temperature stability 9, 9*f*
Crystalline state 6, 7*f*, 35

D
Derived
  defined 35
Devitrify
  defined 35
Diatomite 5, 35
Diatoms 3–5, 4*f*
Dimension stone 15, 35
Displacive transformation 35

E
Egypt 15
Epidemiological studies 35
Ethyl silicate 35

F
Feldspar 11
Fine china 15
Free silica 35
Fused quartz 36
Fused silica 36

G
Glass 15, 17*f*
  crystalline silica brick in
    furnaces 18, 40
Glassy state 6–7, 36. *See also*
  Amorphous state
Gravimetric sampling 36
Grinding wheels 15

H
Hazard Communication Standard vii,
  22–23, 29
HCS. *See* Hazard Communication
  Standard
High quartz. *See* Beta quartz
Host rock 36

I
IARC. *See* International Agency for
  Research on Cancer
Igneous rocks 10–11, 10*f*, 36, 39
  crystalline silica in 10–11
International Agency for Research on
  Cancer 21, 36
  carcinogenic categories 24–25
  carcinogenic classification of
    silica 25–26
  evaluation process 23–25

J
Jasper 36

K
Keatite 9, 36

L
Lithify
  defined 36
Low quartz. *See* Alpha quartz

M
Macrophage 36
Magma 36
Material impairment 36–37
Measurement
  techniques 28*t*
  variables 27, 29
Metal 37
Metalloids 1–2, 37
Metamorphic rocks 10, 10*f*, 12, 37, 39
  quartz in 10
Metastable
  defined 37
Methylsilicate 37
Mine Safety and Health
  Administration 26–27
Mineral 37
Mining operations
  and crystalline silica 12, 13*t*–14*t*
Molecule 37
Museum of Natural History 15

N

National Toxicology Program 23
Neoplasm 37
Noncrystalline silica 8*f*
Noncrystalline state 6, 7*f*, 8, 37
Nonmetal 37
Nonrepeating patterns 6, 7*f*

O

Occupational Safety and Health
   Administration vii, 22–23, 38
Oil and gas industry 18
Opal 37
Optics industry 19
Organic compounds 5, 38
OSHA. *See* Occupational Safety and
   Health Administration
Overburden 38

P

Permissible exposure limit 38
Phagocytize
   defined 38
Piezoelectric quartz crystals vii, 19
Piezoelectricity 38
Pliny 12, 15
Pneumatic hammer drill 21
Polymerization 38
Polymers 5
Polymorph 38
Polymorphism 5
Porcelain 15
Precipitated silica 38

Q

Quartz vii, 2, 6, 8*f*, 9, 20, 38
   alpha 8*f*, 9
   beta 8*f*, 9
   in building materials 15
   crystals 2–3, 3*f*, 5
   in dust 10
   in igneous rocks 10, 11
   in metamorphic rocks 10, 12
   in rocks 10–11
   in sand 10
   in sedimentary rocks 10, 12

in soil 10
temperature stability 8–9, 9*f*

R

Radiolarian earth 38
Radiolarians 3–5
Radiolarite 39
Reconstructive transformation 39
Regulatory agencies. *See* International
   Agency for Research on Cancer,
   Mine Safety and Health
   Administration, National
   Toxicology Program,
   Occupational Safety and Health
   Administration
Respirable crystalline silica 39
Rock cycle 10–11, 10*f*, 39

S

Safe Drinking Water and Toxic
   Enforcement Act (California) 27
Sand vii, 2, 40
Sandblasting 21
Sandpaper 15
Sedimentary rocks 10–11, 10*f*, 12, 39
   quartz in 10, 12
Semiconductors 2, 39–40
Silica 1, 2*f*, 8, 8*f*. *See also* Crystalline
   silica
   amorphous 3, 5
   of biological origin 3
   as chemical compound 2–5
   noncrystalline 8*f*
   relationships with silicon, silicates,
      silicones 1–5, 2*f*, 8, 8*f*
Silica brick 18, 40
Silica flour 15, 40
Silica gel 40
Silica sand 40
Silica W 40
Silicate minerals 40
Silicates 2*f*, 5, 8, 8*f*, 40
Siliceous 40
Silicic rock 40
Silicon 1–2, 2*f*, 8, 8*f*, 40–41
Silicon carbide 41

Silicon dioxide 41
Silicon-oxygen tetrahedra 2-3, 5, 41
Silicon polymers 5
Silicon tetrahydride 41
Silicones 2*f*, 5, 8, 8*f*, 41
Silicosis vii, 21, 28, 41
Smithsonian Castle 15, 17*f*
Stable
    defined 41
States of matter 42
Steel industry 18
Stishovite 42
Synthetic crystalline silica 19

T
Tetrahedra 2-3, 3*f*, 42
Tridymite 8*f*, 11, 20, 42
    temperature stability 8-9, 9*f*
2a substance 33

U
U.S. Occupational Safety and Health
    Administration. *See* Occupational
    Safety and Health Administration
U.S. Treasury building 15

V
Vitreous silica 42
Vitrify
    defined 42
Volcanic tuff 11, 42

W
Washington, George 15
White House 15